Introduction

It is 300 million years ago—long before the first dinosaurs appeared on Earth. You are in a forest, part of which is flooded with shallow, murky water. The rest is so wet that the ground is squishy. The air is muggy, and the whole place is buggy. You're in a swamp!

Cockroaches are climbing trees to escape hungry centipedes. A dragonfly is devouring insects in midair, while millipedes are chomping on rotting plants. And something keeps poking its green head out of the still water. . .

Too bad you can't really jet back in time and see for yourself what the creature is. Maybe it's a *good* thing, because you would be in for a big surprise. Several big surprises. The roaches would be larger than your hand, the centipedes longer than your arm. That dragonfly's wings would stretch as wide as a kite, and one millipede would fill your sleeping bag.

ONE SMALL SQUARE®

SWAMP

by Donald M. Silver

illustrated by Patricia J. Wynne

New York San Francisco Washington, D.C. Auckland Bogotá
Caracas Lisbon London Madrid Mexico City Milan
Montreal New Delhi San Juan Singapore
Sydney Tokyo Toronto

LEARNING
TRIANGLE
PRESS

Connecting kids, parents, and teachers
through learning

An imprint of McGraw-Hill

Every plant and animal pictured in this book can be found with its name on pages 38–43. If you come to a word you don't know or can't pronounce, look for it on pages 44–47. The small diagram of a square on some pages shows the distance above or below the surface for that section of the book.

For Granny Cathey

patient, kind, caring, giving

Sincere thanks to Dr. Robert S. Voss and Darrel Livesay for sharing their swamp experiences with us; and to Marc Gave, Maceo Mitchell, and Thomas L. Cathey for all their efforts on our behalf.

Text copyright © 1997 Donald M. Silver.
Illustrations copyright © 1997 Patricia J. Wynne.
All rights reserved.
One Small Square® is the registered trademark of Donald M. Silver and Patricia J. Wynne.

ISBN 978-0-07-057926-2
MHID 0-07-057926-1
9 10 11 12 LWI 24 23 22

Whether you are in a swamp or at home, always obey safety rules! Neither the publisher nor the author is liable for any damage that may be caused or any injury sustained as a result of doing any of the activities in this book.

Even the forest itself would amaze you. Back then the towering plants were horsetails, ferns, and club mosses—not oaks, maples, and pines.

As for that green-headed creature, Eogyrinus, you wouldn't want to tangle with him. He was as long as a van and might mistake you for dinner!

The ancient swamp is gone forever. So are all its giant-sized plants and animals. But there are still swamps on Earth, full of surprises. If you explore a swamp, you may meet up with trees with "knees", manatees, snappers, and flytrappers. You also could wind up stuck in muck, stranded by floating islands, or just plain lost. That's why you must NEVER explore a swamp alone, ONLY with an experienced guide.

Until then—but still with an adult—you can explore one small square of a swamp from the boardwalk in a park or nature preserve. This book will explain how. If there is no swamp nearby, explore along with this book. There will be activities you can do at home. Then you will be prepared for your first visit to a real swamp.

What's Living in a Swamp?

Monera

Protists

Funguses

Plants

Animals

With an inexpensive magnifying glass and other simple equipment, you will be able to peek and poke into parts of a swamp without harming yourself or anything that lives there.

One Small Square of a Cypress Swamp

What have you heard? That swamps are creepy, swamps are slimy, swamps stink? That if you wander into a swamp, you may never get out—if a snake doesn't get you, a gator will? That if you take one wrong step, the ground will tremble, then you'll sink? Or maybe that zombies run wild at night as hideous monsters rise out of the mist?

The truth is, most of this swamp talk was meant only to give a good scare. But it helped many people believe that swamps were not just scary but useless. Swamp after swamp was destroyed. Swamps that remain today are still in danger.

If only everyone knew how helpful and important swamps are. That's where you come in. If you learn how a swamp works, you can convince others that swamps must be protected and saved.

Here you see one small square of a cypress swamp. On page 21 there is a square of a mangrove swamp. Each square is about 4 feet (1.2 m) long on a side. That's around the size of a four-person elevator. Step up on the boardwalk: A plant is about to eat an animal.

Even in a swamp with a boardwalk, people get lost. So ALWAYS take an adult with you when you explore. Ask what to wear, and what to use to keep biting insects away.

Fly By or Bye Fly

What's sparkling in the small square? If you were on the board-walk, you could kneel down for a closer look. It's a sundew plant with very odd leaves. At the end of their thin stalks they are covered with clear, gooey drops that glisten in the sunlight.

A fly circling a leaf senses that the drops are as sweet as honey. It lands, hungry for a meal, only to find that the drops are as sticky as glue.

The fly tries to free itself, but it is too late. The drops just won't let go. The harder the fly tries to escape, the more drops it touches. Soon the stalks and

Plant leaves have tiny cells that capture energy from sunlight. The cells use this energy to make sugar in the process called photo-synthesis. At the same time they make oxygen that living things breathe. To make foods such as proteins, plants need nitrogen and minerals too.

Leaf cell

It's a good thing that a sundew's flowers are not near the leaves. Trapped insects can't spread pollen needed for making seeds.

All's fair when it comes to get-ting food. A frog and a spider sit, ready to grab insects away from pitcher plants' deadly traps.

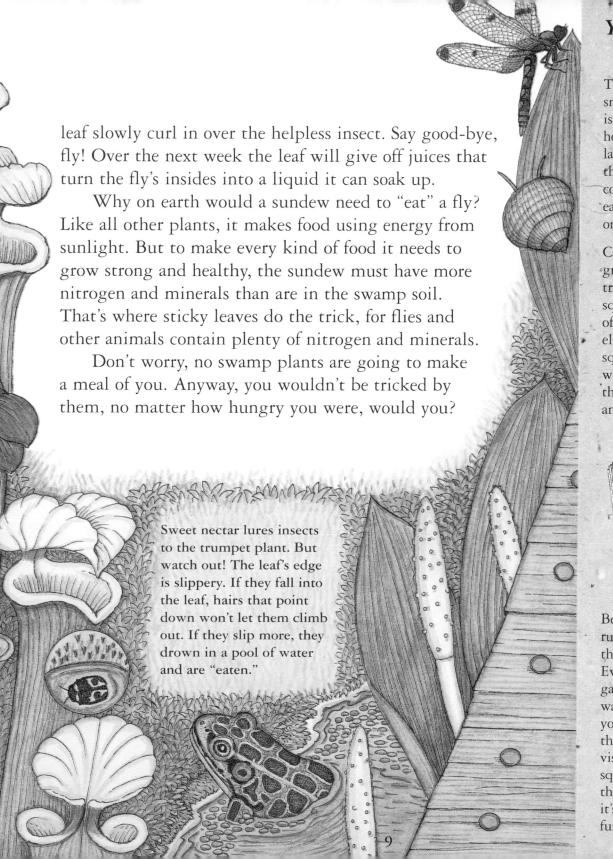

leaf slowly curl in over the helpless insect. Say good-bye, fly! Over the next week the leaf will give off juices that turn the fly's insides into a liquid it can soak up.

Why on earth would a sundew need to "eat" a fly? Like all other plants, it makes food using energy from sunlight. But to make every kind of food it needs to grow strong and healthy, the sundew must have more nitrogen and minerals than are in the swamp soil. That's where sticky leaves do the trick, for flies and other animals contain plenty of nitrogen and minerals.

Don't worry, no swamp plants are going to make a meal of you. Anyway, you wouldn't be tricked by them, no matter how hungry you were, would you?

Sweet nectar lures insects to the trumpet plant. But watch out! The leaf's edge is slippery. If they fall into the leaf, hairs that point down won't let them climb out. If they slip more, they drown in a pool of water and are "eaten."

9

Your Swamp Square and Notebook

The hardest part of choosing a small square from a boardwalk is figuring out which one and how big. Since a swamp is wet land where trees grow, be sure there is a tree in your square. Of course, you will also want insect-eating plants. If you can't find one square with both, try two.

Carry a notebook. Draw a diagram of your square. Include the tree, how many steps wide the square is, how far from the start of the boardwalk, and anything else that will help you find your square again. Fill your notebook with words and pictures about the swamp. Jot down the date and time of day of each visit.

Be sure to follow all swamp rules. Never throw anything in the water. Never feed animals. Even if your swamp has no alligators or poisonous snakes, watch where you sit and what you touch. Remember, you share the boardwalk with other swamp visitors, so let them see your square when they want. And if the adult with you gets bored, it's up to you to show how much fun it is to explore a swamp.

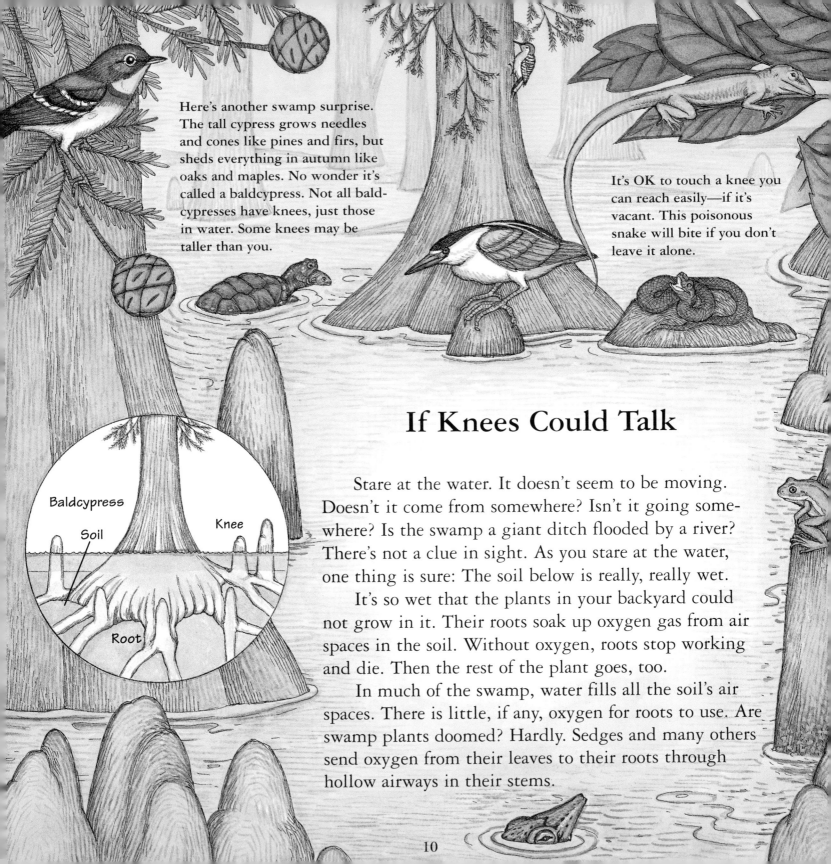

Here's another swamp surprise. The tall cypress grows needles and cones like pines and firs, but sheds everything in autumn like oaks and maples. No wonder it's called a baldcypress. Not all bald-cypresses have knees, just those in water. Some knees may be taller than you.

It's OK to touch a knee you can reach easily—if it's vacant. This poisonous snake will bite if you don't leave it alone.

Baldcypress

Soil

Knee

Root

If Knees Could Talk

Stare at the water. It doesn't seem to be moving. Doesn't it come from somewhere? Isn't it going somewhere? Is the swamp a giant ditch flooded by a river? There's not a clue in sight. As you stare at the water, one thing is sure: The soil below is really, really wet.

It's so wet that the plants in your backyard could not grow in it. Their roots soak up oxygen gas from air spaces in the soil. Without oxygen, roots stop working and die. Then the rest of the plant goes, too.

In much of the swamp, water fills all the soil's air spaces. There is little, if any, oxygen for roots to use. Are swamp plants doomed? Hardly. Sedges and many others send oxygen from their leaves to their roots through hollow airways in their stems.

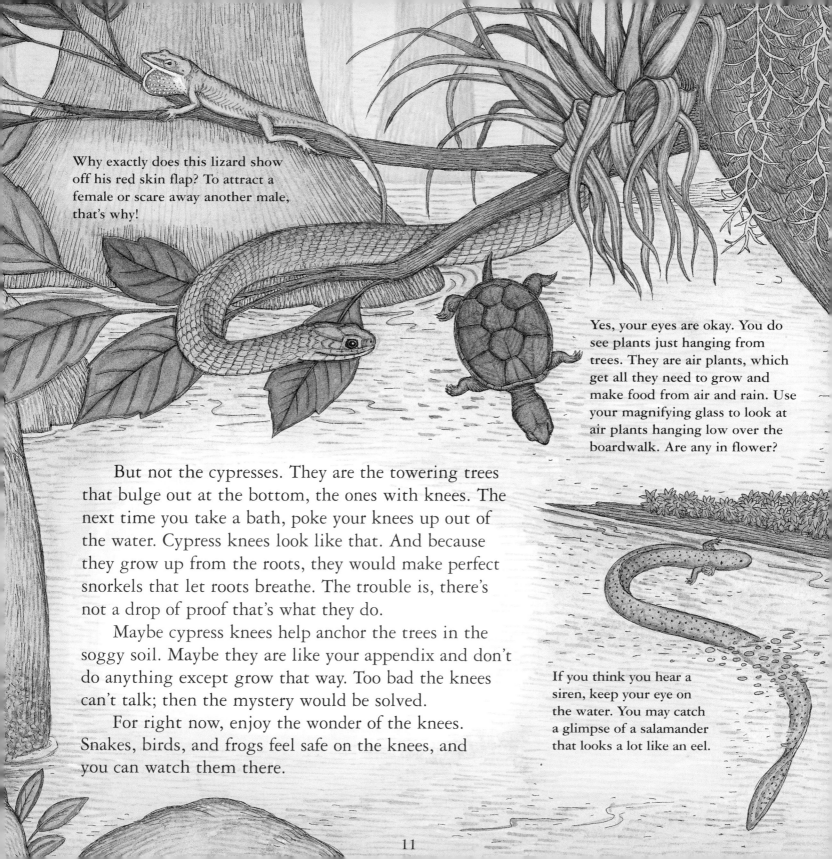

Why exactly does this lizard show off his red skin flap? To attract a female or scare away another male, that's why!

Yes, your eyes are okay. You do see plants just hanging from trees. They are air plants, which get all they need to grow and make food from air and rain. Use your magnifying glass to look at air plants hanging low over the boardwalk. Are any in flower?

But not the cypresses. They are the towering trees that bulge out at the bottom, the ones with knees. The next time you take a bath, poke your knees up out of the water. Cypress knees look like that. And because they grow up from the roots, they would make perfect snorkels that let roots breathe. The trouble is, there's not a drop of proof that's what they do.

Maybe cypress knees help anchor the trees in the soggy soil. Maybe they are like your appendix and don't do anything except grow that way. Too bad the knees can't talk; then the mystery would be solved.

For right now, enjoy the wonder of the knees. Snakes, birds, and frogs feel safe on the knees, and you can watch them there.

If you think you hear a siren, keep your eye on the water. You may catch a glimpse of a salamander that looks a lot like an eel.

Spores

2

Eggs

1

Sperm

3

Ferns make more ferns with spores(l). Spores grow into sperm-and-egg makers(2). A new fern(3) grows after a sperm joins with an egg.

Sphagnum moss

Hollow cell

A carpet of sphagnum moss covers this floating peat island. The moss's tangled leaves have special hollow cells that soak up and hold water. As old moss dies, new moss grows on top of it. Guess what happens to the old moss? It forms more peat.

Bacteria and funguses break down a rotting log into simple nutrients that plants can reuse.

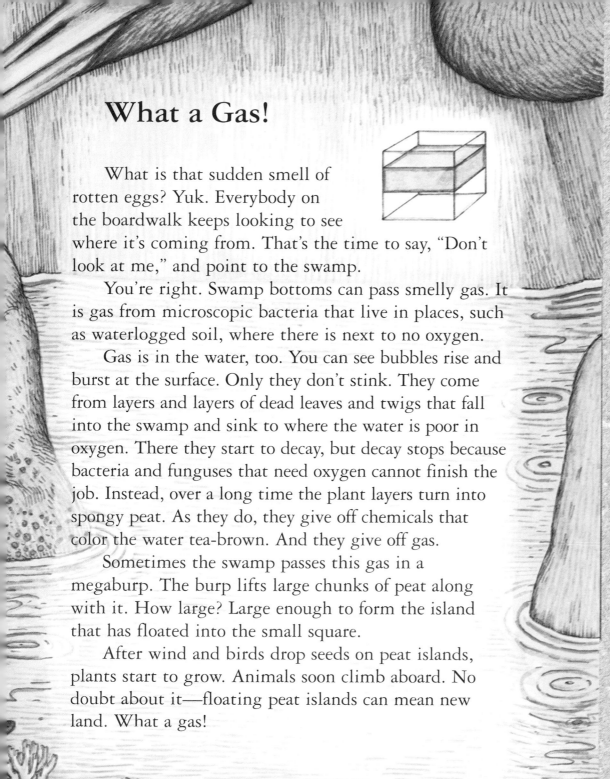

What a Gas!

What is that sudden smell of rotten eggs? Yuk. Everybody on the boardwalk keeps looking to see where it's coming from. That's the time to say, "Don't look at me," and point to the swamp.

You're right. Swamp bottoms can pass smelly gas. It is gas from microscopic bacteria that live in places, such as waterlogged soil, where there is next to no oxygen.

Gas is in the water, too. You can see bubbles rise and burst at the surface. Only they don't stink. They come from layers and layers of dead leaves and twigs that fall into the swamp and sink to where the water is poor in oxygen. There they start to decay, but decay stops because bacteria and funguses that need oxygen cannot finish the job. Instead, over a long time the plant layers turn into spongy peat. As they do, they give off chemicals that color the water tea-brown. And they give off gas.

Sometimes the swamp passes this gas in a megaburp. The burp lifts large chunks of peat along with it. How large? Large enough to form the island that has floated into the small square.

After wind and birds drop seeds on peat islands, plants start to grow. Animals soon climb aboard. No doubt about it—floating peat islands can mean new land. What a gas!

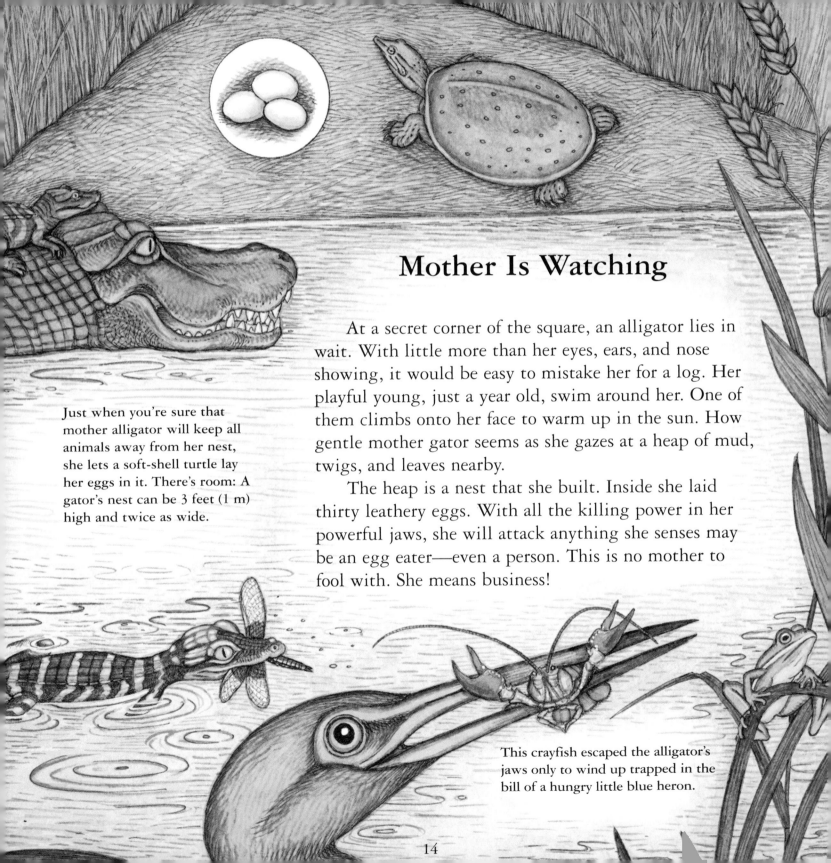

Mother Is Watching

Just when you're sure that mother alligator will keep all animals away from her nest, she lets a soft-shell turtle lay her eggs in it. There's room: A gator's nest can be 3 feet (1 m) high and twice as wide.

At a secret corner of the square, an alligator lies in wait. With little more than her eyes, ears, and nose showing, it would be easy to mistake her for a log. Her playful young, just a year old, swim around her. One of them climbs onto her face to warm up in the sun. How gentle mother gator seems as she gazes at a heap of mud, twigs, and leaves nearby.

The heap is a nest that she built. Inside she laid thirty leathery eggs. With all the killing power in her powerful jaws, she will attack anything she senses may be an egg eater—even a person. This is no mother to fool with. She means business!

This crayfish escaped the alligator's jaws only to wind up trapped in the bill of a hungry little blue heron.

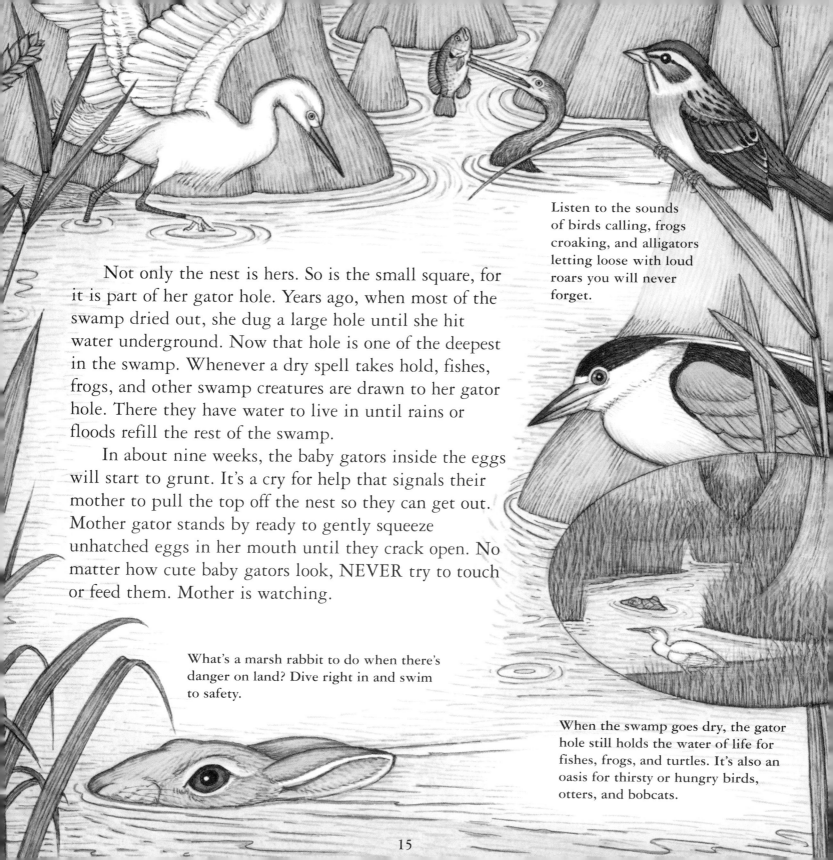

Listen to the sounds of birds calling, frogs croaking, and alligators letting loose with loud roars you will never forget.

Not only the nest is hers. So is the small square, for it is part of her gator hole. Years ago, when most of the swamp dried out, she dug a large hole until she hit water underground. Now that hole is one of the deepest in the swamp. Whenever a dry spell takes hold, fishes, frogs, and other swamp creatures are drawn to her gator hole. There they have water to live in until rains or floods refill the rest of the swamp.

In about nine weeks, the baby gators inside the eggs will start to grunt. It's a cry for help that signals their mother to pull the top off the nest so they can get out. Mother gator stands by ready to gently squeeze unhatched eggs in her mouth until they crack open. No matter how cute baby gators look, NEVER try to touch or feed them. Mother is watching.

What's a marsh rabbit to do when there's danger on land? Dive right in and swim to safety.

When the swamp goes dry, the gator hole still holds the water of life for fishes, frogs, and turtles. It's also an oasis for thirsty or hungry birds, otters, and bobcats.

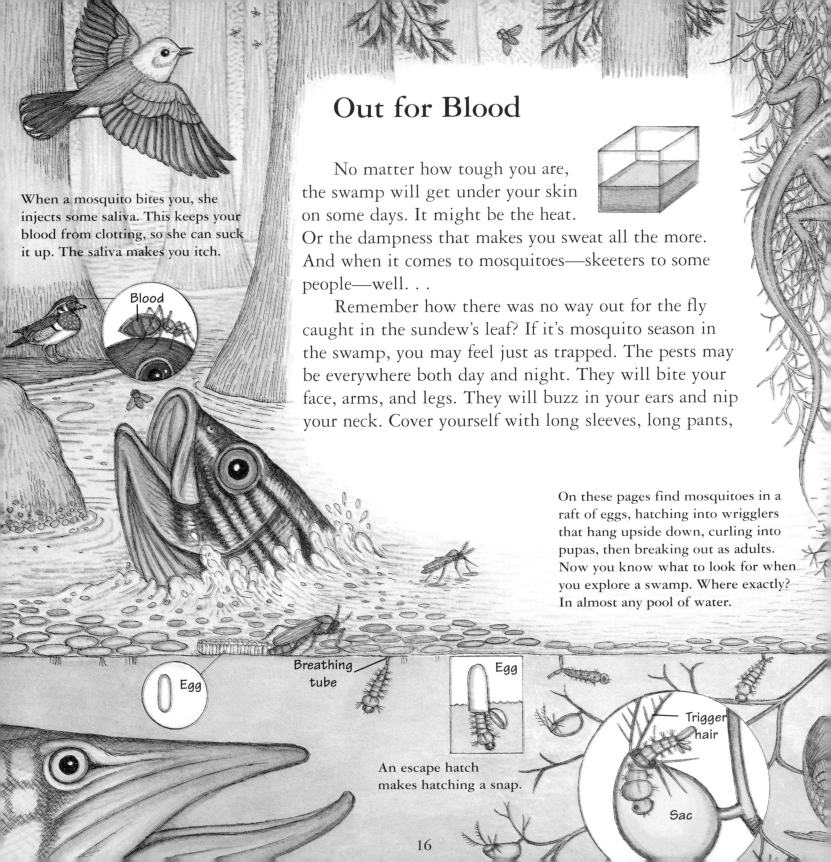

Out for Blood

When a mosquito bites you, she injects some saliva. This keeps your blood from clotting, so she can suck it up. The saliva makes you itch.

Blood

No matter how tough you are, the swamp will get under your skin on some days. It might be the heat. Or the dampness that makes you sweat all the more. And when it comes to mosquitoes—skeeters to some people—well. . .

Remember how there was no way out for the fly caught in the sundew's leaf? If it's mosquito season in the swamp, you may feel just as trapped. The pests may be everywhere both day and night. They will bite your face, arms, and legs. They will buzz in your ears and nip your neck. Cover yourself with long sleeves, long pants,

On these pages find mosquitoes in a raft of eggs, hatching into wrigglers that hang upside down, curling into pupas, then breaking out as adults. Now you know what to look for when you explore a swamp. Where exactly? In almost any pool of water.

Egg

Breathing tube

Egg

An escape hatch makes hatching a snap.

Trigger hair

Sac

16

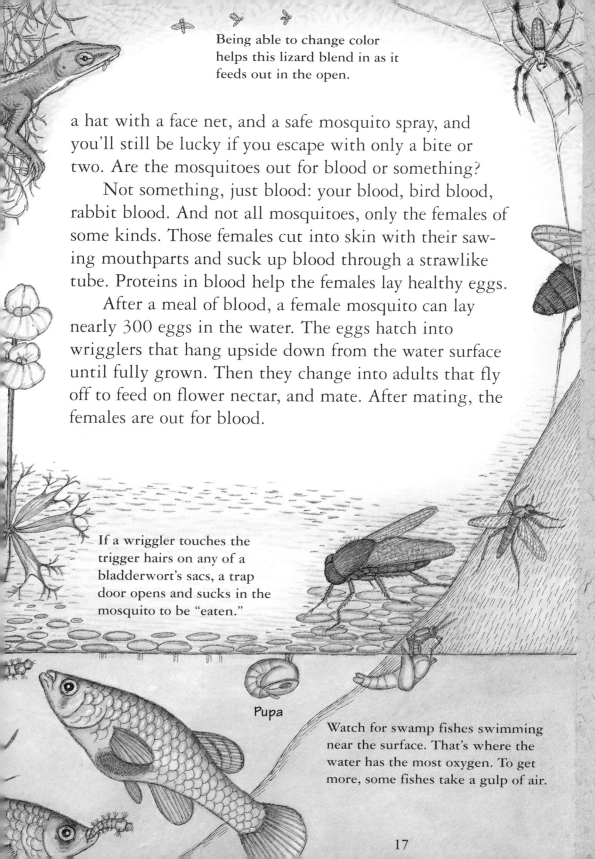

Being able to change color helps this lizard blend in as it feeds out in the open.

a hat with a face net, and a safe mosquito spray, and you'll still be lucky if you escape with only a bite or two. Are the mosquitoes out for blood or something?

Not something, just blood: your blood, bird blood, rabbit blood. And not all mosquitoes, only the females of some kinds. Those females cut into skin with their sawing mouthparts and suck up blood through a strawlike tube. Proteins in blood help the females lay healthy eggs.

After a meal of blood, a female mosquito can lay nearly 300 eggs in the water. The eggs hatch into wrigglers that hang upside down from the water surface until fully grown. Then they change into adults that fly off to feed on flower nectar, and mate. After mating, the females are out for blood.

If a wriggler touches the trigger hairs on any of a bladderwort's sacs, a trap door opens and sucks in the mosquito to be "eaten."

Pupa

Watch for swamp fishes swimming near the surface. That's where the water has the most oxygen. To get more, some fishes take a gulp of air.

17

Body Talk

Without words, people use body talk to signal other people. They smile, frown, blush, stare, cry, growl, shake a fist, etc. Animals do the same. When a poisonous cottonmouth opens its mouth and shows its fangs, the message is clear: Back off!

Start a *Body Talk* list in your notebook. List wordless messages you get from family, friends, teachers, and pets. Anytime you see an animal in the backyard, park, woods, or swamp, note its body talk and what you think it means. Was the message meant for another animal or for you?

COUGAR   BOBCAT

RABBIT   MOOSE   RACCOON

Leaving a Mark

Have you ever left footprints in mud or snow? Animals leave their mark all the time. Draw footprints you find in your yard or in a park and try to figure out what made them.

Look for prints in the muddy earth on any land you walk through to reach the boardwalk and on any land in your square. If an alligator's been by, you may discover a trail of tail marks, too. Draw what you see. Do any footprints or marks match these?

At night bats locate moths and mosquitoes—their food—by bouncing high-pitched sounds off them.

If you hear a cry that sounds like "Who-cooks-for-you," there's a barred owl living in the swamp. With tasty mice to catch, it needs no cook.

Newborn opossums are the size of rice. They crawl into their mother's pouch to drink her milk and grow. Soon they are big enough to ride on her back.

Pouch

Often there will be no warning that a mosquito is about to attack. Should you hear the buzz of her wings, you might be quick enough to swat her before she gets you. But when the bumps on your skin start to itch, you may find yourself out for blood, wishing that someone would just get rid of all mosquitoes.

Wrong, wrong, wrong. The swamp is the mosquitoes' home, not yours! It is where they fit in and make a living. Getting rid of all mosquitoes means killing the ones that never bite people. The swamp would suffer because mosquitoes are food for many a swamp animal.

Small mosquitofishes, for instance, eat wrigglers. Gars and other large fishes prey on the small fishes and, in turn, may become an alligator's dinner. Birds, frogs, dragonflies, and bats hunt adult mosquitoes—the ones that bite people *and* the ones that don't. When these mosquito eaters are well fed and have young, there's plenty of food for other swamp predators such as snapping turtles, snakes, bobcats, and of course, alligators.

Predators are out for blood because they must eat to stay alive. So don't blame mosquitoes for doing what they must to keep their kind alive. Maybe one day you'll discover a way to stop mosquitoes from biting people without harming them or their swamps.

Hunters you never see by day prowl the swamp at night. As long as predators are hungry, prey must beware.

One Small Square
of a Mangrove Swamp

Don't rush to put away your mosquito protection. You'll surely need it when you visit a mangrove swamp, especially if the sand flies are out in force with the mosquitoes. Just try swatting these biting flies. They are so small they are nicknamed "no-see-ums."

You can expect a mangrove swamp to be hot, damp, and often smelly. So why leave the alligators? Because mangrove swamps are too incredible to miss.

Look at the small square shown here. It's the same size as the square of cypress swamp, but the two are so different. This square is at the edge of the ocean. The shallow water is salty, with waves and tides, and the creatures are mostly those of the sea. A tangle of roots makes it nearly impossible for anyone to simply put on a pair of boots and wade very far.

That's why the best way to explore a mangrove swamp is by canoe—and ONLY with an adult guide who knows the swamp. For now, come explore from the boardwalk of the small square in this book. You'll quickly discover what surprises a mangrove swamp has in store.

MANATEE
ZONE

NO BOATING

Before you step onto the boardwalk, be sure to read all the rules that are posted. They are there to make sure that both you and the swamp are never harmed.

Red mangrove flower

While still on the tree, the seed inside a mangrove fruit grows into an icicle-shaped seedling. If the seedling lands in the mud, a new tree may grow there. If the seedling floats away, it may take root somewhere else. No matter where a red mangrove grows, its roots are able to absorb seawater but keep out most of the salt. The leaves release what little salt the roots do absorb.

Fruit

Air enters

For red mangroves to grow, the water must be shallow. And it takes prop roots to anchor the trees in the wet, slippery mud.

Seedling

Look up and you may spot a mangrove cuckoo hunting spiders or a snake sunning on a branch. Look down and search for fishes.

The Swamp Makers

Did you ever hear of trees that "give birth"? Trees that make new land? Trees that move into the sea? Well, meet red mangroves, the trees in the square.

The first time you see a red mangrove, you may catch it giving birth. Just look for something falling from a branch that is yellow-green, icicle-shaped, and about as long as this book. It's a mangrove seedling that has started growing while still attached to its "parent" tree. If the pointed end of the seedling lands in the mud, it will send down roots. Later, food-making leaves will sprout from the other end.

Once in the mud, mangrove roots, like cypress roots, need oxygen. The wet mud has hardly any. But it's no mystery how they get the oxygen. As mangroves grow, prop roots arch out of their trunks. They curve down into the water and disappear in the mud. On each prop root small white spots above the waterline take in air for the rest of the root below.

The tangle of prop roots traps mud and fallen leaves, twigs, and bark. These form a thick mass that starts to decay, but then very slowly turns into muddy peat— new land for more mangroves and for more prop roots to invade the sea.

Prop It Up

Do mangroves really need prop roots to help hold them up? Try standing a pen or an unsharpened pencil straight up on its flat end. What happens? Wrap one end of a pipe cleaner around the pen. Arch the pipe cleaner like a prop root and fold it as shown. How many pipe cleaners does it take to get the pen to stand? What happens if you add more and more? How hard does it become to knock over the pen?

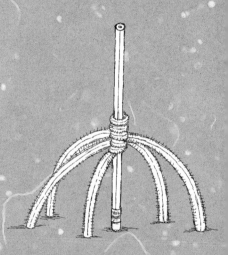

A-maze-ing

It takes a raccoon to climb over the tangle of prop roots in a mangrove swamp. Or a crab to crawl under. From the boardwalk it is hard to tell where prop roots start and where they end. Try drawing the tangle of prop roots in your small square. Later, at home, draw a maze of prop roots and try it on your family and friends. Once in the maze, can they ever get out?

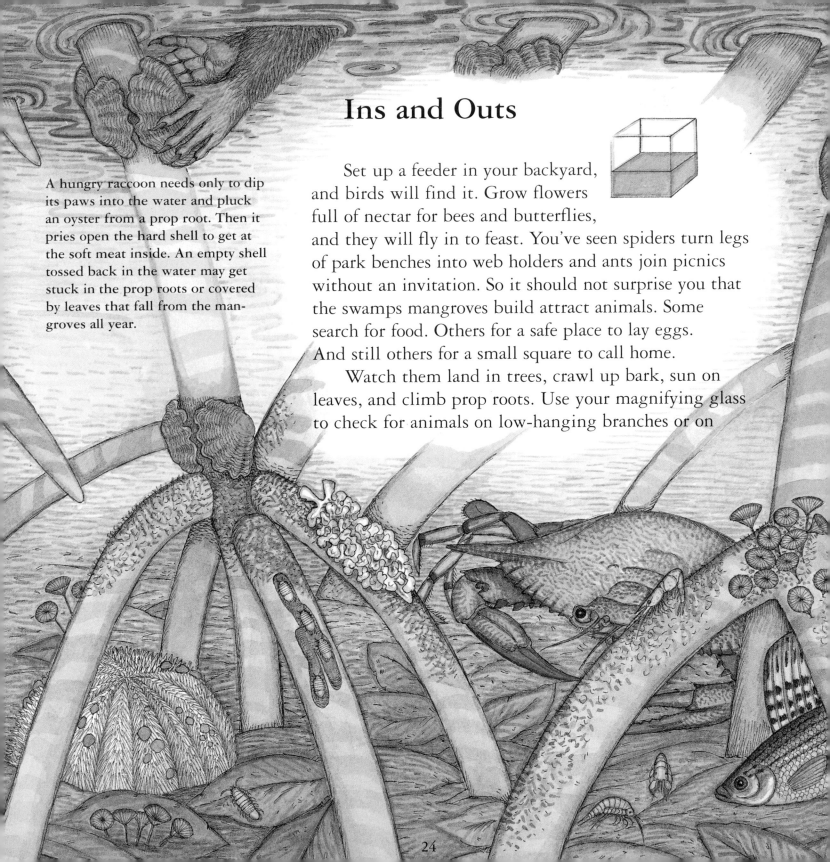

Ins and Outs

A hungry raccoon needs only to dip its paws into the water and pluck an oyster from a prop root. Then it pries open the hard shell to get at the soft meat inside. An empty shell tossed back in the water may get stuck in the prop roots or covered by leaves that fall from the mangroves all year.

Set up a feeder in your backyard, and birds will find it. Grow flowers full of nectar for bees and butterflies, and they will fly in to feast. You've seen spiders turn legs of park benches into web holders and ants join picnics without an invitation. So it should not surprise you that the swamps mangroves build attract animals. Some search for food. Others for a safe place to lay eggs. And still others for a small square to call home.

Watch them land in trees, crawl up bark, sun on leaves, and climb prop roots. Use your magnifying glass to check for animals on low-hanging branches or on

prop roots that touch the boardwalk. There may be snails or crabs right in front of your eyes that you could miss unless they start moving.

Just when you think you've spotted a creature, it may disappear into the tangle of prop roots. Don't give up. Prop roots are great for hiding because they act as roadblocks that few large predators can get past, from land or sea.

For every animal you see above the surface of the water there are hundreds below, busy making a living. At high tide shrimps, worms, and small fishes nibble on fallen mangrove leaves. Oyster shells open to take in seawater that contains oxygen to breathe and bits of food to

Algae are food makers too. They float in water with tiny creatures that eat them. The algae and the creatures together are known as plankton. Algae also grow on prop roots.

The legs-up dance of a male blue crab invites a female to mate. When a blue crab hatches from its egg(1), it grows(2), sheds its shell, and develops five pairs of legs(3).

Alga

Plankton

Match Game

Whether you are exploring a cypress or a mangrove swamp, you will come across plants and animals you have never seen before. To help you identify them, you will need a field guide. This book contains names and pictures of animals and plants that make their homes in different places. For example, there are separate guides for trees, birds, seashells, and reptiles. But there are also habitat guides that contain many of the living things you may discover in one place, such as a swamp. You may have field guides at home. If not, you can find them at the library. When you are out exploring, draw what you see. Make notes. If a leaf is as big as your hand, write that down. If it is light-green, jot that down too. Later compare your drawings to the pictures and descriptions in the field guides. You'll be surprised how often you can make a match.

eat. While one blue crab dances to attract a mate, another hunts a small fish.

At low tide most of the small square is exposed to the air. Fishes and shrimps must swim to the edge, where some water stays when the tide goes out. If they don't, they will dry out and die from lack of oxygen. Oysters can stay put if they shut their shells tightly with enough water inside to breathe until the tide returns. Worms just dig down into the wet mud, where they can breathe and stay moist.

Any animal that remains out in the open had better have ways of protecting itself. Low tide signals fiddler crabs to crawl out of their burrows and browse the mud for food. Hungry birds that may have missed a meal under the water have little trouble spotting their prey. A hard shell or colors that blend into the mud can mean the difference between life and death.

Take advantage of low tide to really observe the prop roots. What's living on them? What's stuck in them? Are any stepping out into the sea, the site of future swamp?

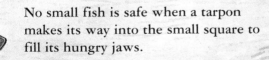

No small fish is safe when a tarpon makes its way into the small square to fill its hungry jaws.

Low tide is the ideal time to witness body talk. That's when male fiddler crabs wave their colorful large claw at other males to go away or at females to come by.

Thousands of baby fishes dart among mangrove roots. Why shouldn't they? It's their nursery. Here they find plenty to eat, and they avoid many predators kept away by shallow water and tangled roots.

How hard is it to eat without hands? Watch the many ways birds use their beaks to spear, pull, or filter a meal out of the water. Maybe you'll even see an anhinga toss a fish in the air and swallow it head first.

Spoonbill

Brown pelican

Anhinga

Take a guess how many birds are in a mangrove tree at nesting time. Then count. There may be over fifty!

For the Birds

Get your binoculars ready. Maybe your earplugs, too—in case the loud chorus of grunts, squawks, cheeps, and wheeps isn't music to your ears. For when it comes to life out of water, mangroves are truly for the birds.

Here, high and low, on branches and on prop roots, birds sleep, sun, and build nests. From the boardwalk it is easy to admire what expert builders they are. If it looks simple to you, just try making a nest out of mud, twigs, and leaves the next time you are in a park.

Just because a baby bird outgrows its nest, doesn't mean it can fly or find its own food. For the time being, it must wait for its parents to spit up fish and other food they swallowed.

One way to quiet baby birds is to keep the food coming.

No matter how close a nest is to you, don't touch it. There may be eggs or even chicks inside. And birds will protect their young from harm as fiercely as any mother alligator. After all, that's what the noise is all about. As baby birds cry out to be fed, their parents warn other birds to BACK OFF and GET LOST or they'll be sorry.

But there's another reason never to upset a nest in a mangrove tree. It might belong to a brown pelican or a wood stork. These birds are endangered. If they don't keep having lots of babies, they may soon die out. So do your share and leave the mangroves for the birds.

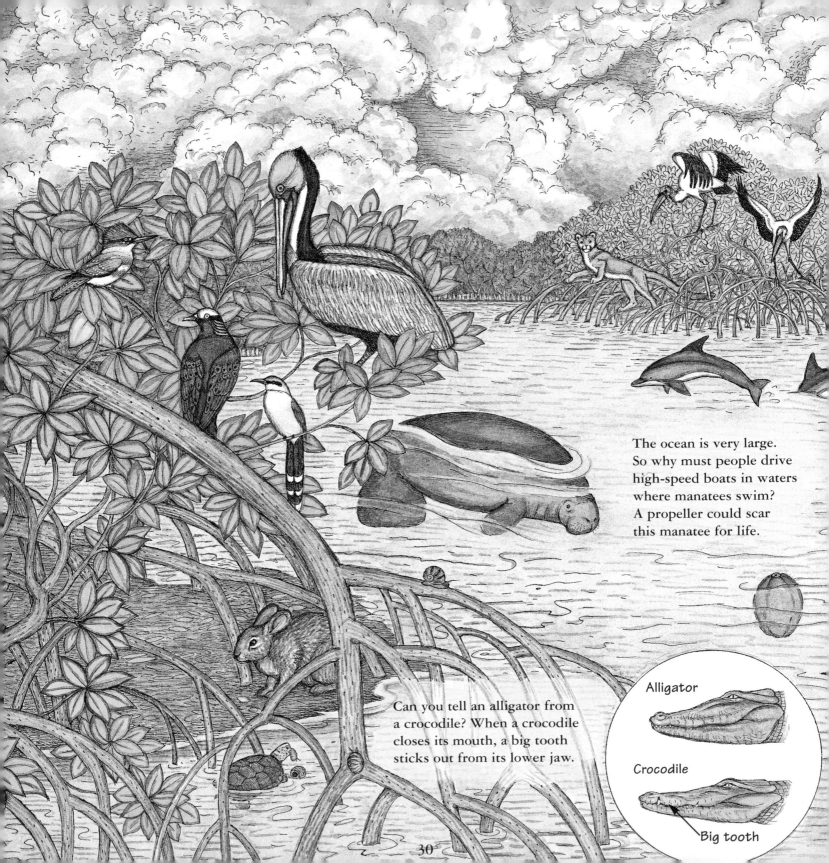

The ocean is very large.
So why must people drive
high-speed boats in waters
where manatees swim?
A propeller could scar
this manatee for life.

Can you tell an alligator from
a crocodile? When a crocodile
closes its mouth, a big tooth
sticks out from its lower jaw.

Alligator

Crocodile

Big tooth

The mangrove trees' march into the sea is slow going. First they must help build the land on which their prop roots can take hold and their seedlings can grow. Year after year they push out into the ocean a little at a time.

Saved by the Swamp

Did you see the dark clouds gathering in the distance? Or feel the breeze begin to build into a wind? Well, somebody did, because the boardwalk is emptying fast. There's a big storm brewing. Unless you want to get soaked, you better seek shelter too.

But not before you take a quick last look at the small square. There's a chance you'll glimpse a crocodile, a turtle, a deer, or a manatee seeking a safe place to wait out the storm. Why in the swamp? Because mangroves are very strong. Their trunks, branches, and roots break the force of winds and waves that otherwise could harm the animals.

About the only storms mangroves can't withstand are really powerful hurricanes. Their giant waves and violent winds can topple almost an entire swamp. Storms, though, also wash in mangrove seedlings. Before you know it, there's a new tangle of prop roots on the way.

Seconds Count

When storms come up, early warning can pay off. Swamp animals can sense danger and turn to mangroves for shelter. When a tornado warning comes over the television, people run to safety. Nature gives you a thunderstorm warning. When you see lightning, count how many seconds before you hear thunder. If you count 5, the storm is about a mile (1.6 km) away. If 10, double that. Heed the warning and find a safe place. NEVER stand under a tree, for lightning may strike it.

Hurricane Who?

Every year tropical storms form over the ocean and build into powerful hurricanes. They are given names such as Gloria, Hugo, and Dora. Listen to weather broadcasts and read newspapers during hurricane season to track each storm. In your notebook keep a record of wind speed, wave heights, and where each hurricane hits land—if it does. If a hurricane strikes your town, write your own account of what it's like to face nature's fury.

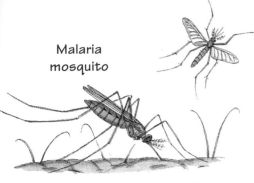

Malaria mosquito

In many parts of the world, mosquitoes are deadly. They spread fatal diseases such as malaria to the people they bite. The germs are in the saliva they inject with the bite. For now, the only way to try to protect people is to spray whole swamps to kill mosquitoes and their eggs. Scientists are trying to chemically change the mosquitoes so that the deadly germs can't live inside them and can't be passed along in a bite.

Can you match each living thing with its outline?

Not only animals are saved by swamps. So are homes of families that live where rivers and streams flood after big storms. When the floodwaters reach a nearby swamp, the swamp slows them down, soaks them up, and holds them. The damage to homes is much less. Afterward, the swamp releases the water a little at a time through the soil. The soil acts as a filter, removing harmful chemicals the water picked up during the flood. The water becomes underground water, safe enough to drink.

If you hear people say that the swamps should be drained and all the trees cut down to build homes and farms, you now know what to tell them. That gators and other swamp life have nowhere else to go. That baby fishes and birds grow up there. That swamps are home to endangered wildlife. That they make new land. And that they often save animals and homes during storms.

The more you explore your small square, the more you can tell people about how important swamps are. Spreading the word really helps. Already there are laws in many places to protect swamps and other wetlands (see page 36). Who knows? One day you might become a swamp guide.

Red mangroves do all the work of building land but can't live totally out of water. Where the land made by the mangroves rises above the water, red mangroves start to die off and black mangroves start to grow in their place.

Carolina anolis

Upside-down jellyfish

Variegated urchin

Sheepshead minnow

Belted
kingfisher

llow-rumped
warbler

Red mangrove

White
ibis

Mangrove
cuckoo

Blue crab

Mud fiddler
crabs

Crown conch

Mangrove
diamondback
terrapin

Sea
slug

Land hermit
crab

Mullet

Pink shrimp

Bubble melampus

Amphipods

Virgin nerite

Grey snapper

Flat tree oyster

Sailfin
molly
and
mosquito

Mangrove
tunicate

Frons
oyster

33

Swamp in a Box

Measure the length and height of a shoebox. Cut a piece of paper for the background wall about 1/4 inch (6 mm) shorter than the box height and about 4 inches (10 cm) longer than its length. Draw cypress or mangrove trees on it. Place the picture in the box and tape each side to the front. The picture will curve.

On separate sheets draw and color swamp plants and animals, each with a flap. Decide where the flap should be by where you want to glue or tape the plant or animal. Cut out each picture, bend its flap, and glue or tape it in place.

Create one box for a mangrove swamp at low tide. Create another for a cypress swamp with a floating island. Use pipe cleaners for mangrove prop roots and cut a piece off a kitchen sponge for the floating island.

Hand in Hand

If you like fossils, you will love this: Swamps and fossils often go hand in hand. For hundreds of millions of years, some of the best fossils ever found formed in swamps.

A fossil is anything left behind by an animal or plant that lived a very long time ago. Bones can be fossils. So can teeth or shells. Or the outline of a leaf pressed into a rock. Fossils are how we know that giant cockroaches, giant ferns, and Eogyrinus all lived in ancient swamps.

When most living things died in ancient swamps, they were eaten or they rotted away without leaving a trace. But some plants that fell into swamp waters didn't fully rot because there was hardly any oxygen. They formed thick layers of peat. As new layers pressed on older ones, they squeezed water out of the lower layers and created heat. Floods left layers of sand and soil to press the peat even more. Over thousands of years the peat hardened into coal.

If you look at pieces of coal—a "fossil fuel"—under your magnifying glass, you can often see bits of leaves, wood, and bark. They are fossils left by plants that grew in swamps 250 to 400 million years ago.

The peat in your small square may become coal. And, in the far future, someone may dig it up and wish to rocket back in time to get a look at ancient creatures like you.

One Small Square of a Cypress Swamp

Can you match each living thing with its outline?

Spanish moss

Carolina anolis

Elisa skimmer dragonfly

Grass pink

Trumpet plant

Swamp sparrow

Baldcypress

White peacock butterfly

Parula warbler

Kingbird

Swamp rose

Sphagnum moss

Red-bellied woodpecker

American alligator

Prothonotary warbler and mosquito

Rough green snake

Warmouth bass

Emetic russala

Two-lined salamander

Black-crowned night heron

Mosquito-fishes

Mystery snail

Pickerel frog and tadpole

Green sunfish

Crayfish

Long-nosed gar

Duckweed

Wood duck

Yellow unicorn entobama

Lesser siren

Mud turtle

35

Wetlands, NOT Wastelands

Do you know a place covered by shallow water for part of the year? Or a spot with soggy soil for months at a time? Then you know a wetland! Wetlands aren't wastelands. All wetlands are important for plants, animals, and people in the same ways swamps are.

The trees in this northern swamp are mostly maples. Instead of peat, the soil is mineral-rich like that of most woods. Neither alligators nor crocodiles live here, but many birds, deer, and salamanders do.

Open land that is flooded by rivers or lakes is known as a slough. One may be within driving distance of your home.

Marshes are scattered all over the place. They are home to grasses, reeds, and cattails—but no trees. Some are freshwater, others saltwater. If a pond near your house is filling in with mud and plants, it is turning into a marsh.

Bogs are wetlands where you may see no water. But don't be fooled. The land is peat covered by sphagnum moss. Both are water-logged. They may tremble and give way. Bogs soak up water only from rain or melting snow. Boglike areas with some water from a river are called fens. Decay takes place so slowly in bogs that scientists were able to study part of the brain of a human being buried in a Florida bog more than 7,000 years ago.

Peat

About 10,000 years ago, thick ice covered this land. When it melted, it scraped out very large holes. Today, when these prairie potholes fill with water they are great places to find many kinds of ducks.

To the Rescue

Can you imagine! Not too long ago alligators were in danger of becoming extinct. Today they are doing just fine because of laws that protected them from harm. But many, many other wetland plants and animals are still endangered, such as those you see here. Rescuing them starts with saving their wetland homes.

Swamp pink

Florida panther

Reed parrotbill

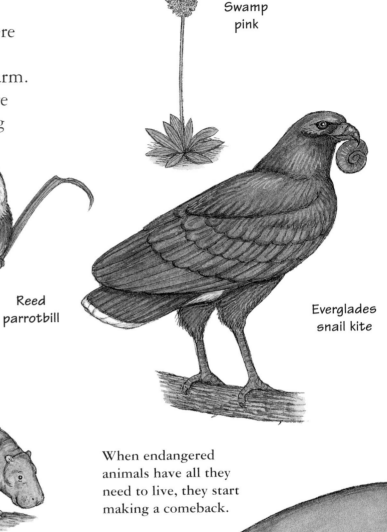

Everglades snail kite

Saving endangered wildlife means protecting them from hunters.

Pigmy hippopotamus

When endangered animals have all they need to live, they start making a comeback.

Manatee

Swamp deer

One way to save water life is to stop water pollution.

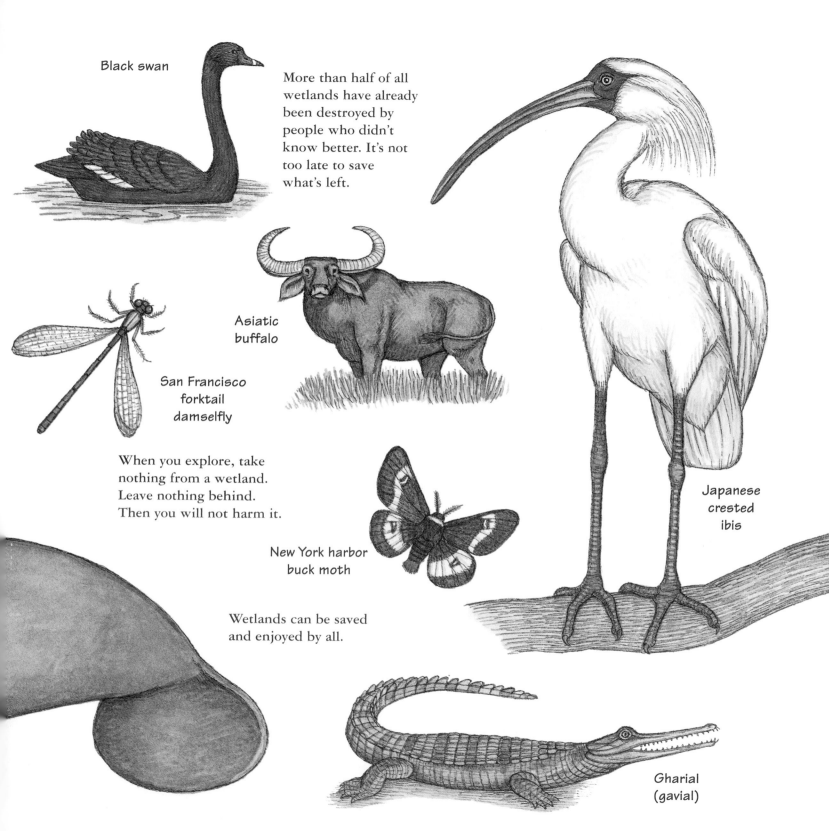

Black swan

More than half of all wetlands have already been destroyed by people who didn't know better. It's not too late to save what's left.

Asiatic buffalo

San Francisco forktail damselfly

When you explore, take nothing from a wetland. Leave nothing behind. Then you will not harm it.

New York harbor buck moth

Wetlands can be saved and enjoyed by all.

Japanese crested ibis

Gharial (gavial)

What gives swamps their air of mystery? Maybe it's the food makers: cypress trees, mangroves, sundews, and all the other plants shown here. Mushrooms and other funguses often look like plants, but they are unable to make food. Some swamp animals are invertebrates. They don't have a bone in their bodies. Look for them in the air, in the soil, on plants, and in the water.

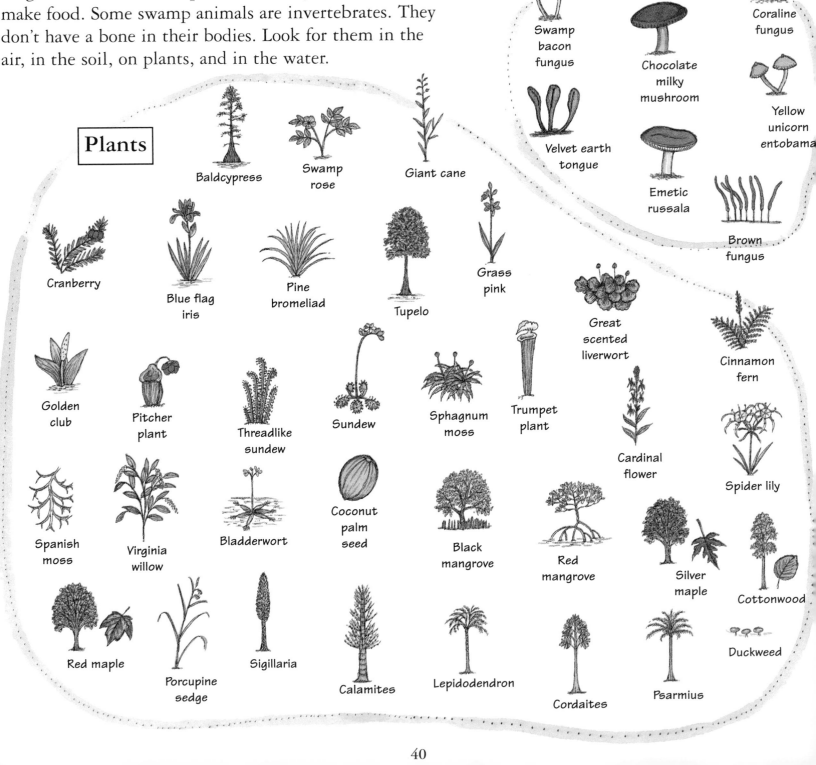

Funguses

Swamp bacon fungus

Velvet earth tongue

Chocolate milky mushroom

Emetic russala

Coraline fungus

Yellow unicorn entobama

Brown fungus

Plants

Baldcypress

Swamp rose

Giant cane

Cranberry

Blue flag iris

Pine bromeliad

Tupelo

Grass pink

Great scented liverwort

Cinnamon fern

Golden club

Pitcher plant

Threadlike sundew

Sundew

Sphagnum moss

Trumpet plant

Cardinal flower

Spider lily

Spanish moss

Virginia willow

Bladderwort

Coconut palm seed

Black mangrove

Red mangrove

Silver maple

Cottonwood

Red maple

Porcupine sedge

Sigillaria

Calamites

Lepidodendron

Cordaites

Psarmius

Duckweed

Invertebrates

Ancient millipede

Ancient centipede

Giant dragonfly

Ancient spider

Ancient springtail

Onychophora

Culex mosquito

Malaria mosquito

Giant cockroach

Primitive insect

Vinegar fly

Mud fiddler crab

Flat tree oyster

Frons oyster

Upside-down jellyfish

Green darner

Elisa skimmer dragonfly

Palamades swallowtail

Widow dragonfly

Mystery snail

Golden silk spider and eggs

Swamp milkweed beetle

White peacock butterfly

Crayfish

Colona moth

Mangrove skipper caterpillar and adult

Land hermit crab

Virgin nerite

Arrow worm

Sea gooseberry

Blue crab and larva

Crown conch

Isopod

Amphipod

Variegated urchin

Pink shrimp

Mangrove tunicates

Sand fly

Polychaete worm

Sea slug

Bubble melampus snail

Grass shrimp

American horsefly

Little vase sponge

Sheet-web weaver spider

41

All these swamp animals have backbones. They are vertebrates. If you see an animal with fur, it's a mammal. With feathers, it's a bird. Most fishes and reptiles have scaly skin, while an amphibian's skin is moist and scaleless.

Fishes

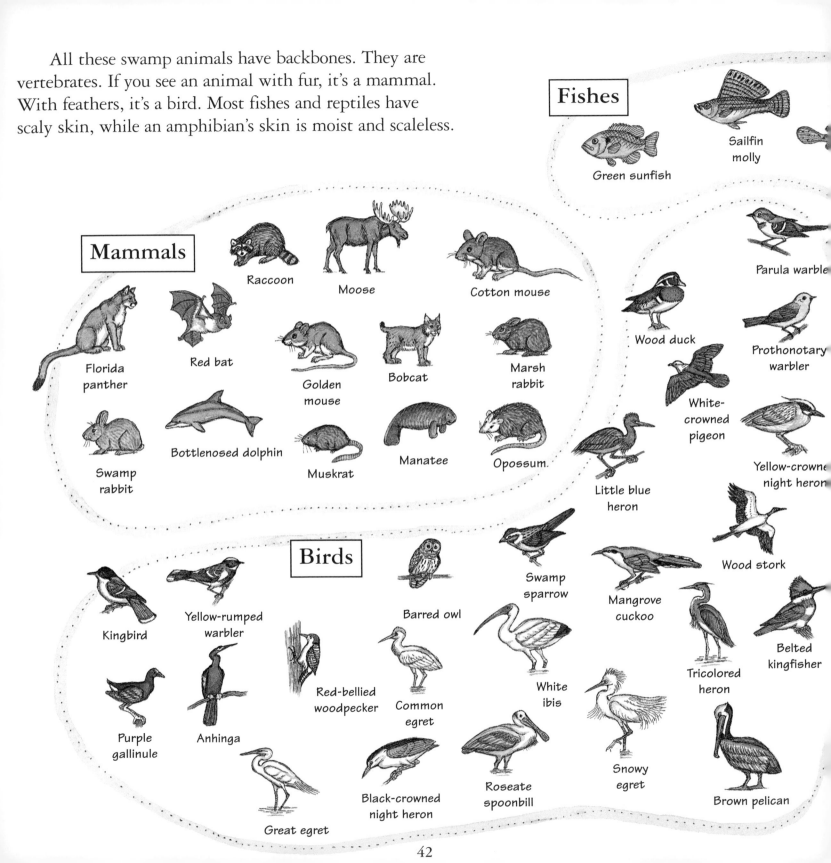

Green sunfish

Sailfin molly

Mammals

Raccoon

Moose

Cotton mouse

Parula warbler

Florida panther

Red bat

Golden mouse

Bobcat

Marsh rabbit

Wood duck

Prothonotary warbler

Swamp rabbit

Bottlenosed dolphin

Muskrat

Manatee

Opossum

White-crowned pigeon

Yellow-crowned night heron

Little blue heron

Wood stork

Birds

Barred owl

Swamp sparrow

Mangrove cuckoo

Tricolored heron

Belted kingfisher

Kingbird

Yellow-rumped warbler

Red-bellied woodpecker

Common egret

White ibis

Purple gallinule

Anhinga

Black-crowned night heron

Roseate spoonbill

Snowy egret

Brown pelican

Great egret

 Tarpon

 Sea horse

 Long-nosed gar

 Diamond killifish

 Mullet

Mosquitofish

Warmouth bass

 Grey snapper

Sergeant major

 Sheepshead minnow

 Golden shiner

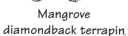 Chain pickerel

Reptiles and Amphibians

 Common snapping turtle

Bird-voiced tree frog

 American alligator

American crocodile

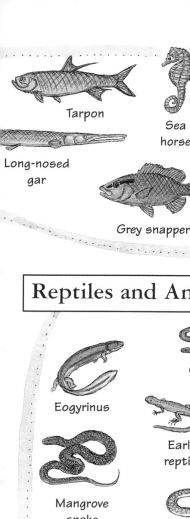

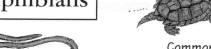 Ophiderpeton

Eogyrinus

 Early reptile

 Mud turtle

 Two-lined salamander

 Carolina anolis

Mangrove diamondback terrapin

Mangrove snake

 Rough green snake

Though you need a microscope to see protists and monera, you may smell gases some of them give off in the swamp.

Lesser siren

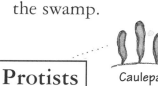 Soft-shelled turtle

Protists

Caulepa

Diatom

 Callithamnion

 Mermaid's cup

Halimeda

Monera

Green tree frog

Cottonmouth

 Pickerel frog

 Red algae

 Shelled ameba

 Bacteria

Index

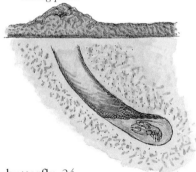

Index

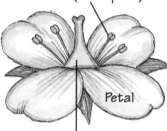

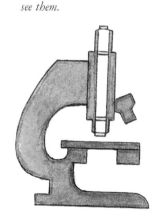

Index

mineral 8, 9. *A kind of chemical that, in living things, helps a cell work the way it should. Calcium and iron are two examples of minerals.*

monera (muh-NEER-uh) 5, 43. *Creatures made up of one cell that doesn't have a nucleus (control center).*

mosquito 16, 17, 18, 20, 32, 41

mosquitofish 18

moss 12. *See also* **sphagnum moss**

moth 18

mouse 18

mouth 29

mouthpart 17

mud 14, 22, 23, 26, 28, 36

mushroom 40

N

nectar 9, 17, 24. *Sweet liquid that flowers make to attract bees and other pollen carriers.*

needle 10

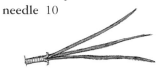

nest 14, 15, 28, 29

nitrogen 8, 9. *A gas in the air that some soil bacteria can change into a form plants can absorb and use to make proteins.*

northern swamp 36

nose 14

nutrient (NOO-tree-int) 12. *Any part of food that living things must have to build cells or to use as a source of energy.*

O

oak 5, 10

ocean (*see* **sea**)

Okefenokee swamp 13

opossum 18

otter 15

owl 18

oxygen (AHK-suh-jin) 8, 10, 13, 17, 23, 25, 26, 34. *A gas in the air that living things breathe.*

oyster 24, 25, 26

P

paw 24

peat 12, 13, 23, 34, 36, 37

photosynthesis (foh-toh-SIN-thuh-siss) 8. *Process by which plants and some protists use light energy from the sun to make food out of water and carbon dioxide gas.*

pine 5, 10

pitcher plant 8

plankton 25

pollen (PAHL-in) 8. *Powdery grains, made by flowers, that contain male reproductive cells.*

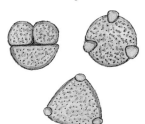

pollution 38

pond 36

pouch 18

prairie pothole 37

predator (PRED-uh-tur) 18, 19, 25, 27. *Animal that kills other animals for food.*

prey 19, 26. *Animal hunted or caught for food by a predator.*

prop root 22, 23, 24, 25, 28, 31, 34

protein (PRO-teen) 8, 17. *Substance that cells must make to grow and do the chemical work of staying alive and healthy. Proteins contain nitrogen.*

protist 5, 43. *A living thing usually of one cell that has a nucleus (control center).*

pupa (PYOO-puh) 16. *Part of the life cycle of some insects in which their bodies change completely as they become adults.*

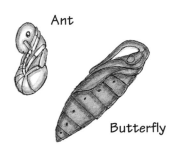

Ant

Butterfly

R

rabbit 15

raccoon 23, 24

rain 11, 15, 37

reed 36

reptile 26, 42

river 10, 32, 36, 37

rock 34

root 10, 11, 20, 22, 23, 27, 31. *See also* **prop root**

rotting 34. *See also* **decay**

Index

Fish **Butterfly**

Reptile

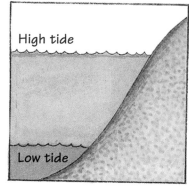

High tide

Low tide

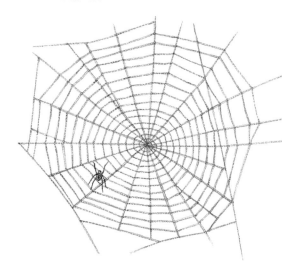

Find Out More

To learn more about swamps and other wetlands write to:

Environmental Protection Agency

Office of Wetlands Protection

Office of Water

Washington, DC 20460

Cache River State

National Area

930 Sunflower Lane

Belknap, IL 62908

Further Reading

Look for the following in a library or bookstore:

Golden Guides, Golden Press, New York, NY

Golden Field Guides, Golden Press, New York, NY

The Audubon Society Beginner Guides, Random House, New York, NY

The Audubon Society Field Guides, Alfred A. Knopf, New York, NY

The Peterson Field Guides, Houghton Mifflin Co., Boston, MA

Reader's Digest North American Wildlife, Reader's Digest, Pleasantville, NY

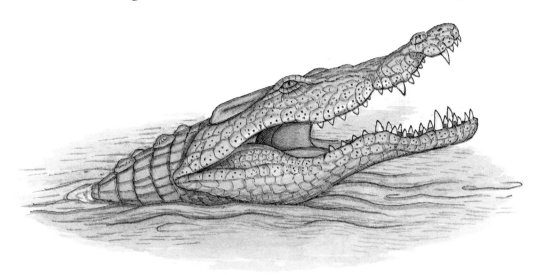